LOCAL HEROES

PORTRAITS OF AMERICAN VOLUNTEER FIREFIGHTERS

4880 Lower Valley Road • Atglen, PA 19310

Photographs by IAN SPANIER
Text by MAREK FUCHS
Design by GRACE MARTINEZ and FLORIAN BACHLEDA

Other Schiffer Books on Related Subjects:
Aerial Firefighting, 978-0-7643-3068-1, $39.99
German Firefighting Vehicles in World War II, 0-7643-0191-8, $29.95

Library of Congress Control Number: 2012938498

Design by Grace Martinez & Florian Bachleda
Type set in Jannon/Knockout/MillerDisplay/MillerText/Times

ISBN: 978-0-7643-4150-2
Printed in China

Schiffer Books are available at special discounts for bulk purchases for sales promotions or premiums. Special editions, including personalized covers, corporate imprints, and excerpts can be created in large quantities for special needs. For more information contact the publisher:

Published by Schiffer Publishing, Ltd.
4880 Lower Valley Road
Atglen, PA 19310
Phone: (610) 593-1777; Fax: (610) 593-2002
E-mail: Info@schifferbooks.com

For the largest selection of fine reference books on this and related subjects, please visit our website at:
www.schifferbooks.com
We are always looking for people to write books on new and related subjects. If you have an idea for a book, please contact us at proposals@schifferbooks.com

This book may be purchased from the publisher.
Please try your bookstore first.
You may write for a free catalog.

In Europe, Schiffer books are distributed by
Bushwood Books
6 Marksbury Ave.
Kew Gardens
Surrey TW9 4JF England
Phone: 44 (0) 20 8392 8585; Fax: 44 (0) 20 8392 9876
E-mail: info@bushwoodbooks.co.uk
Website: www.bushwoodbooks.co.uk

CONTENTS

FOREWORD

FAITH IN AMERICAN INSTITUTIONS, it is safe to say, is at a low ebb in the second decade of the twenty-first century. Yet the smaller and more local you go, the institutions that remain closest to their communities are as revered as ever. Chief among them are our country's volunteer firefighting crews, whose life and work we chronicle here.

Volunteer firefighting, after all, is a story of small towns and outsize personalities, of bake sales and barbecues, all punctuated by the blood-draining exhilaration of risk and rescue. As children, we dream of gearing up to slide down a fire-station pole and jump aboard an engine as, with a racing wail, it sets off toward distress.

This instinctive intrigue does not fizzle once we grow up. Adulthood, though, brings with it the responsibilities of work and family. Yet some never relinquish this dream — an undeniably good thing. Few municipalities other than large cities can finance a paid firefighting force, so a considerable majority of firefighters in America are volunteers - who in turn account for about half of all line-of-duty deaths every year.

Volunteer firefighting, in fact, is a vocation as old as the nation itself. Though often assumed to be lost to time and the headlong rush of the modern age, it remains the centerpiece of many American communities.

What would compel a retired teacher, a lawyer, an industrial designer, a pharmacist, an arborist and even an Emmy-nominated songwriter to devote the better part of their spare time and take harrowing risks to protect their neighbors as well as total strangers? We wanted to show these too-often-hidden heroes, stalwarts of their communities, as the dedicated citizens they are: determined but compassionate, weathered but vulnerable, common but singular.

Initially, we photographed the firefighters on a stark white background, creating an image as much about the subject as about the larger figure they represent. As the project progressed, we added some environment to the shots. We approached the lighting the same but now added elements of their surroundings — from the familiar comfort of the fire station to the dark husks of burned-out mountainsides where lives had been lost.

For those photographed here, even decades of service will end not with a flush pension but rather a commemorative plaque and a brief (if heartfelt) round of applause. Yet through day and night, billows of snow in the North, humid nights in the South, the thin air of the high mountainside and the battered cliffs of the Pacific, these men and women - in an act as selfless as it is timeless - drop their lives and rush out to help. And so we set off on a journey across America, visiting its most captivating volunteer firehouses and meeting the local heroes inside.

—*Ian, Marek and Florian*

FIREFIGHTING 101

*T*HE CALL CAN COME at any time — and it usually does. It might be midnight or in the middle of a lazy summer afternoon or in the still water calm of the break of dawn. The volunteer firefighter — he or she could be a dad, a mom, a recent high school dropout, a wealthy bond salesman, or a retired plumber — is going about their normal routine when the call goes out... In years past, and still in some small towns, a loud siren from the town center would wail, reverberating for miles. Today, though, in this age of high technology, most often the tone of a pager sounds at home or at the office.

However the alert is delivered, the message is the same: Somewhere within the community's confines, there's a report of a fire. Unless it's a false alarm, and there are many of those, some neighbor or parcel of precious land is in trouble. Work or dinner or sleep or a favorite TV show is forgotten, left behind, as the firefighter dashes to the firehouse, where he or she hops on a 30-ton rig and, at breakneck speed, barrels to the scene of the blaze.

THERE ARE 1.15 MILLION FIREFIGHTERS in the United States; nearly three-quarters of them are volunteers, according to the National Fire Protection Association. That means that the vast majority of those risking their lives to fight fires in the United States are not working for the village or town or municipality; for their efforts as first responders to all manner of disasters, they do not receive — or, for that matter, even ask for — one thin dime. In any given year, half of the nation's firefighters who die on the job are volunteers.

Volunteer firefighting would seem to be a relic of lost America, a throwback to a quaint, idyllic time in the popular imagination when small towns were populated by people who would do anything to help a neighbor. There is a widely held assumption, perhaps rooted in the rush and narcissism of modern life, that volunteer firefighting is an anachronism. With the pace of modern life, few really have time to be a firefighter anymore. Right?

Not quite. There's something about volunteer firefighting that captures the spirit of a large group of people, no matter how busy life has become. For that reason, volunteer firefighting is, as the numbers demonstrate, thriving — and that's a lucky thing. Firefighting is a numbers game. To tame most fires, a large team of firefighters is needed, each performing specific tasks and rotating into the fire and then out before their air tanks deplete.

Commonly, only sizeable cities can afford to pay. For the rest, the financial burden would be untenable.

THE HISTORY OF VOLUNTEER FIREFIGHTING in the United States is entwined with that of both firefighting at large and the nation as a whole. Benjamin Franklin and George Washington were volunteer firefighters; the first president of the United States was a member of the Friendship Veteran's Fire Engine Company of Alexandria, Virginia. Even the blistering fires at America's first settlement in Jamestown were fought by a volunteer contingent. In many communities, volunteer fire departments were the first formal community organizations. Though there have clearly been advances since the days of bucket brigades and horse-drawn hose carts, that basic act of ditching the standard workweek job or weekend moment to run to the assistance of a neighbor or a complete stranger has not changed.

Today, wealthier suburban areas might be blessed with state-of-the-art fire engines and equipment updated for the asking. Rural firehouses in poorer regions of the country, by contrast, might have to make do with outdated rigs. They all but beg money with a cup — often running "Fill-the-Boot" donation campaigns at traffic intersections — in an attempt to replace their frayed equipment. Yet across the United States, there are commonalities among volunteer firehouses. Most are filled with big personalities, adventure seekers, and community activists who like their activism to be tinged with risk. However, each volunteer firehouse is also its own ecosystem and subculture, an amalgamation of the peculiarities of the local culture and whatever other strange sets of interactions and behaviors emerge when a group of people spend a little too much time together, occasionally in peril.

Here's a little known fact: Most of the calls answered by volunteer firefighters are uneventful. About 80 percent are false alarms. The other 20 percent are wild cards.

Fires can tamp down quickly or they can be tenacious, resisting every effort to squelch them. The act of fighting a fire is a blend of tactics and preparation, harnessed to snap judgments. Even the most orderly plan to attack a fire can turn into chaos in a hurry, as firefighters, in their aggressive push to help, clog stairwells, even as fire bursts unexpectedly from the landing. In the case of wildfires, communications over vast expanses of land can break down, just as fire jumps the road. Any misstep can spell doom. Even success can at times be the product of luck. Will that beam somehow hold? Or the wind fortuitously shift?

When the fire does finally dissipate, though, volunteer firefighters go back to the routine of their lives — as they have since before the founding of the nation, never a penny richer for their efforts.

HASTINGS HOUSE

HASTINGS-ON-HUDSON, NEW YORK RIVERVIEW MANOR HOSE CO. #3

WHEN AN ATTACHED ROW of wooden walk-up apartment buildings went up in flames in the ice-cloaked early morning hours of winter in 2004, an all-volunteer force of firefighters worked tenaciously though the night, dousing the buildings with thousands of gallons of water to suppress the hard-biting fire, even as their hands froze. When a warehouse ignited several years before that, this same force battled the fierce radiating heat for nearly a dozen hours, until all was safe. Before and since, the volunteers of Riverview Manor Hose Company No. 3 have responded to thousands of emergencies small and large, from kitchen fires to serious automobile accidents.

Most volunteer firehouses in America are, at least to some extent, a cross-section of the local population. There is only one way to get the job — ask! — and no social class, race, or profession has the market cornered on either charity or courage. That makes Hastings-on-Hudson, which sits along the banks of the Hudson River about twenty minutes north of New York City, a particularly interesting case. Historically, Hastings was simultaneously the home of rough-hewn factory workers and those accomplished in the arts, including writers, painters, and play producers.

Factory workers spent long shifts on the production lines of companies like Anaconda Wire and Cable or Zinsser and Company, which produced munitions and mustard gas and used the river for easy transport. Just east of the waterfront, up the hill a bit, lived the likes of Florenz Ziegfeld, the Broadway producer, and his wife Billie Burke, who played "Glinda The Good Witch" in *The Wizard of Oz*, as well as others with credentials and clout aplenty in the arts, attracted by the aesthetic, expansive river views and proximity to the city.

Out of this mix came Riverview Manor Hose Company No. 3, where as diverse a lot as probably exists in firehouses has always been thrown together in close quarters. A composite firefighter here? Impossible. The characteristics, talent, and varied beliefs would know no bounds. Reporting for duty over the company's history have been production line workers, Emmy-nominated musicians, Pulitzer Prize winning journalist Vermont Royster, who also won the Presidential Medal of Freedom, factory foremen, and a poet who fought fires to elegant verse. In many ways, this firehouse represents a social engineering project gone horribly right.

It is hard to say what, if anything, such breadth adds in specific terms to the task of firefighting. The firefighters demure, saying it only makes for a culture that dials down the importance of professional differences. At a fire scene, the wealthiest man in town might very well be answering to the poorest or the most artistic and slight in stature might very well be issuing orders to a brutish looking man who loads flatbeds.

Whatever their store of experience or talents outside the firehouse, members of the Riverview Manor Hose Company have always been bound together by a shared commitment to help their fellow citizens. On September 11, 2001, the company was dispatched to New York City to cover areas of the city left without protection by the terrorist attacks on the World Trade Center.

THE COMPANY WAS BORN in 1910, when a group of local residents wanted their neighborhood prepared for any contingency and were willing to devote their time, skills, and energy to see it get done. They were granted $50 from a neighborhood association and given a hose cart, also known as a "jumper," by the Livingston Hose Company in nearby Dobbs Ferry. Even at the time, the "jumper" was a period piece, but from the moment it arrived, whenever called upon, Riverview Manor has pitched forward to help.

Its inaugural call — for a fire at a coal dock down at the riverfront — came almost immediately. The "jumper" reportedly made it in five minutes, an accomplishment that comes with an asterisk: Coming from the high ground of Hastings, the "jumper" had the advantage of barreling downhill to the scene.

However, the "jumper" did not stand the test of time. Neither did the uniforms of white duck suits and stark white shoes, which the newly formed fire company sported and *The New York Times*, in reporting on a 1915 fire, described as "spotless." Around that time, the founders of Riverview Manor ratified bylaws that meted out considerable fines for slight infractions like drinking, also disallowing "unsavory" figures from spending time in the firehouse.

The company soon bought a used 1908 Locomobile, which came complete with ladders on the side. It also towed a hose cart and a firehouse was built in 1916 to hold the contraption. The house was designed by architect — and Riverview Manor volunteer firefighter — Richmond Shreve, who went on to design the Empire State Building. The house cost $3,550 to build and a fire gong was mounted on the lawn, hammered to sound the call to action.

The Locomobile was succeeded by a 1923 Bull Dog Mac AC 500 Hose Truck. The 1943 Mac Pumper replaced the '23 and served until it was updated with a 1964 Seagrave. Complete with an open cab and crew benches over the hose bed, the '64 Seagrave often ferried twenty firefighters into action.

Decades later, the truck room needed to be expanded once more to house the durable 1984 Haan pumper and a bit of symmetry came into play: A fireman-architect, Captain Bill Switzer, designed the added space.

The next addition to the house came in 2008, to fit the new Seagrave, the latest word in stalwart pumpers, which stretched six feet longer than the Hahn. The truck room addition was completed by George Capuano, a former captain who worked professionally as a stone mason and gave his unending time and scrupulous attention to detail to the expansion. He died only weeks after its completion. Capuano was also the company's principle driver and pump operator. His unexpected death from a heart attack left an emotional tear, but also a large practical void. Who would drive the rig to nearly 200 calls a year? That slack was picked up without question by one of the stone mason's closest friends in the firehouse, Steve Horelick, a musical composer who was twice nominated for an Emmy award.

Charles "Jerry" Saeger

CHARLES "JERRY" SAEGER, who died soon after standing for this photograph, was a poster boy for whiskey and cigarettes, as he lived on the stuff with spunk and spit well into his eighties. Gruff and stubborn, Saeger turned in more years of distinguished service as captain and later chairman, an administrative post, than anyone could easily count. There were jokes around the firehouse after he passed away that Saeger should be propped up on the fire engine like El Cid, leading everyone off into battle one last time. Saeger, a World War II veteran of both the European and Pacific theatres, was also a Dodgers fan. When the Dodgers left Brooklyn for Los Angeles, he waged a fruitless — if steadfast — protest: He refused to set foot inside Yankee Stadium, the home of the last remaining New York baseball team from the Dodgers era in the city, and an arch rival of his beloved and departed Dodgers. However, in 2008, just before the old Yankee Stadium closed, when a fellow firefighter offered a free ticket, Saeger surprised everyone — even himself — by saying, "Yeah, I'll go." At that game, at the age of 84, Saeger pushed aside his beer to catch a foul ball.

VOLUNTEER FIREHOUSES ARE fortunate to have people who do so many different types of day jobs, bringing their expertise to fighting fires. Take Brett Schneiderman, an arborist by trade, who knows as much as anybody about the ins and outs of trees and chain saws. That knowledge often comes in handy and he doles it out to his fellow firefighters, who seek him out in the field and during regular tutorials in the firehouse. He can draw up the perfect response on the subject of trees and offers regular tutorials on chain saw usage. The intellectual rubric of a volunteer firehouse is simple: If you can add a teaspoon of expertise from your profession, please do — and so Schneiderman is the prophet of all things trees and chain saws. When there is a tree appearing to put electrical wires or a home in harm's way, Schneiderman will weigh in, helping everyone find their bearings.

Brett Schneiderman

John Lindner

Paul Faraone

Andrew Sherman

IT WAS THE SECOND NIGHT of Rosh Hashanah, the Jewish New Year, when Andrew Sherman, who would have normally spent the evening with family, heard the call. There was a house fire on Hillside Avenue. He had only been on the force a few months and this would be his first real "worker," as a full-force fire is called. It didn't matter that it was a holy day or that he lacked experience: Sherman barged into the blaze on one of the initial hose lines. The fire was eventually darkened and there were no injuries, but the destructive force of the fire was "pure amazement," Sherman said. "You know that fire is a powerful force," but nothing can truly prepare you to face it. Approaching the house that night, he felt the magnetic energy of the fire and its radiating heat, which only intensified as he entered the house. "Being in a working fire was like witnessing a tsunami," he said. "It gives you a whole new perspective on the force of nature."

AS A JOURNALIST, Marek Fuchs responded to emergencies in Westchester County, New York, to write about them, but he always had to stand behind the line of demarcation; first responders here, everyone else back there. He became a firefighter in Hastings to finally cross the line. After countless calls over many years, Fuchs responded to one in the middle of a wintry storm, a freezing counterpoint to the comfort of the warm bed that he had left behind. As the wind blew and the cold penetrated his bones and as the emergency lights flashed and sirens blared, Fuchs turned toward fellow firefighter Jason Fein and said, "The novelty of this just never wears off."

Marek Fuchs

Dan Hevia

Tony Wan

Chris DiBenedetto

HIS SHOULDERS AND forearms tend to be broader and hands more weathered than the rest of the firefighters. There's good reason: Chris DiBenedetto's day job is among the most physically taxing of all occupations. DiBenedetto works as a firefighter in the Greenwich Village section of New York City. When the workday is done, DiBenedetto comes home to Hastings, where he relaxes with his hobby: Volunteering to fight fires. Though not formally an officer, DiBenedetto runs training sessions in Hastings, handing out hard-earned tips on outmaneuvering smoke, fire, and collapsing structures and pointing out where one wrong move can undermine safety. For the Hastings squad, dedicated but with professional responsibilities far afield of the New York Fire Department, he's an essential connection to the best firefighting expertise in the world. That's why when DiBenedetto gives out praise, his fellow volunteers, who would never show it, soak it in.

JASON FEIN IS A toymaker who became a firefighter six years ago and since then has handled everything from "fires to car extractions — really, everything except cats in trees." There is an obvious and direct boyish connection between toys and the tasks performed in the fire service, but at 42, the link went beyond superhero fantasies or the gleaming, speeding engines. Fein, who owns Guidecraft, an international manufacturer of high-end wooden toys, said that in both fields — firefighting and toys — despite everything you learn in training, you succeed in the world at-large by an intangible instinct.

"It's a leap of faith," he said, about releasing a new product or rushing into a burning building. "Before you were a firefighter, your instinct was to save yourself and run away. Once you are trained, you run right in. Is the difference your superhero costume? Your training?" The toy-maker turned firefighter shook his head. "It's probably, in the end, that simple leap of faith, that simple act of following your heart."

Jason Fein

Lary Greiner

ONE YEAR, LARY GREINER, a mapmaker who answers to the nickname "Bull," answered every call in Hastings, more than three hundred of them. In another year, though, in 2001, in the wake of the 9/11 attacks, he responded to perhaps the single, most important summons of his career. When a paid fire department in the Bronx lost men, Greiner stepped in — and up — to help by joining the force. Whether there was going to be another terrorist attack, as many expected, or a series of run-of-the-mill calls in the Bronx, which were always anything but routine, Greiner knew that this was going to be an essential if backbreaking experience. However, when he first got there, there was time to amble around the firehouse. Lockers were filled with family photos of smiling wives and kids. These men had fully expected to return to them when they went on their last call to the Twin Towers. Greiner was transfixed by the photos of these men who had died and stood ready to answer their call.

Karl "Corky" Soderstrom

Mike O'Toole

WHEN A PAGER BEEPS in the middle of the night, the response "81 en route" comes so rapidly that there are suspicions — only slightly tongue-in-cheek — that Chief James Sarfaty sleeps in his boots and jacket, his walkie-talkie pressed to his ear. Sarfaty, whose shorthand as chief is "81," comes from a long line of volunteer firefighters. His father was chief in nearby Larchmont and his grandfather was a volunteer firefighter too. Sarfaty is the general manager of Atlas Welding and Boiler repair in New Rochelle and, like any chief, stands outside the fire to gain perspective, directing the attack from a slight distance. His judgment is impeccable, according to those he orders about, but standing outside the action makes him fidget. It may not be for him. "It kills him not to be in there," said one of his crew.

Chief James Sarfaty

Steve Horelick

STEVE HORELICK, an Emmy nominated music composer, drives the fire engine, which means he operates the pump panel during fires, controlling how much water runs into the hoses. In the firehouse ecosystem, it is a job of indescribable importance. If he turns a knob the wrong way, knocking water down to a trickle, the firefighters at the other end of the line will have nothing to fight the fire with, which could mean their death. The last time he was nominated for an Emmy, Horelick's fellow firefighters threatened to arrive at the festivities in full gear, complete with air-packs, to cheer him on. It was a joke (much to Horelick's relief since his two "worlds" have never met), but he does see parallels between mixing music and tending to fire scene water flow. In each realm, he works with inputs and outputs on a panel, searching for that magical balance. "Fortunately," said Horelick, "when it comes to mixing music, no one is going to get seriously hurt when the lead guitar isn't quite loud enough."

JENNIFER LEE, who grew up with a pet chimpanzee named Nim Chimpsky, a homage to the linguist Noam Chomsky, and worked for fourteen years at the Bronx Zoo, where she helped design the Congo Gorilla Forest, concentrates on ambulance calls. She is drawn to the challenge of steadying a person in the middle of an excruciating experience, comforting them and tending to their momentary medical needs long enough to reach the hospital. Professionally, Lee doesn't much like to sit still either. She currently travels to different inner-city schools teaching landscape design. When she is not doing that or volunteering to teach stage lighting at the local middle school — or reminiscing about her chimp — her life is punctuated by those frantic emergency calls. Her greatest fear, she said, is sitting "stuck in a windowless office somewhere, in a job I hate."

Jennifer Lee

Michael Adair

John J. Huelsman

DELIGHTED TO BE THE father of three
beautiful daughters, Robert Cadoux's
gratifying and idyllic home life was lacking
only one thing — testosterone — and
so surprising everyone who knew him,
even himself, Cadoux joined the local
fire department. There are, to be certain,
women in local fire departments all across
the nation, but even in modern times, the
numbers are still fairly thin. Cadoux took
right to it: Driving the big engine, handling
the tools, and taking advantage of the easy
marks at the firehouse poker table.

Cadoux, a lawyer by trade, was
particularly taken by the precise nature
of the science of fire suppression. Where
there was a problem at a fire scene, there
was often a corresponding solution based
in the intricacies and complexities of fire
science that he found so compelling. After
a decade, the dad who was just looking
to get his minimum daily requirement of
testosterone had risen to captain.

"If you had ever told me I'd be captain
of a firehouse," he said, when his two-year
term in charge was ending, "I'd have called
you crazy or worse."

Robert Cadoux

Nick Frascone

WITH HIS ARMLOAD of old jokes, Nick Frascone, the open-hearted janitor at the local high school, serves a dual role at the firehouse. Durably built, Frascone will hurtle his body into the crevice of any building or down any embankment. Also, as a born jester, Frascone provides badly needed humor — even when the stakes are highest during a fire or the group is simply waiting around for a call. One of his best bits is to turn any circumstance into a country song, though you might have to plug your ears if he tries to sing it. His only regret in life, he said, is that he never fulfilled his manifest destiny as "an Italian ice salesman on a hot beach." Frascone has many comical mottos that fit in perfectly with the tenor of the humor at a firehouse. His fondest: "Instant asshole, just add water."

DANIEL MANSDORF, always a big boy gently inclined, was a permanent fixture around the firehouse growing up. His unwavering ambition was to become a firefighter. Nearly the moment the clock struck midnight on his 18th birthday, Mansdorf officially became a member, distinguishing himself at the Christmas Eve fire at the local high school soon afterward. Still, as a volunteer firefighter, Mansdorf knew he was on the under-card; he wanted to be a full-time fireman somewhere. He had a friend on the squad in Charleston, South Carolina, a former Riverview Manor firefighter. Charleston needed to replenish its ranks after a fire at a sofa warehouse killed nine firefighters. Mansdorf got the job and, at the age of 20, left Hastings to attend the fire-training academy. When he graduated, a crew of his fellow volunteers flew down to watch the boy who had grown up around the firehouse graduate to the big time.

Daniel Mansdorf

Sudhana Bajracharya

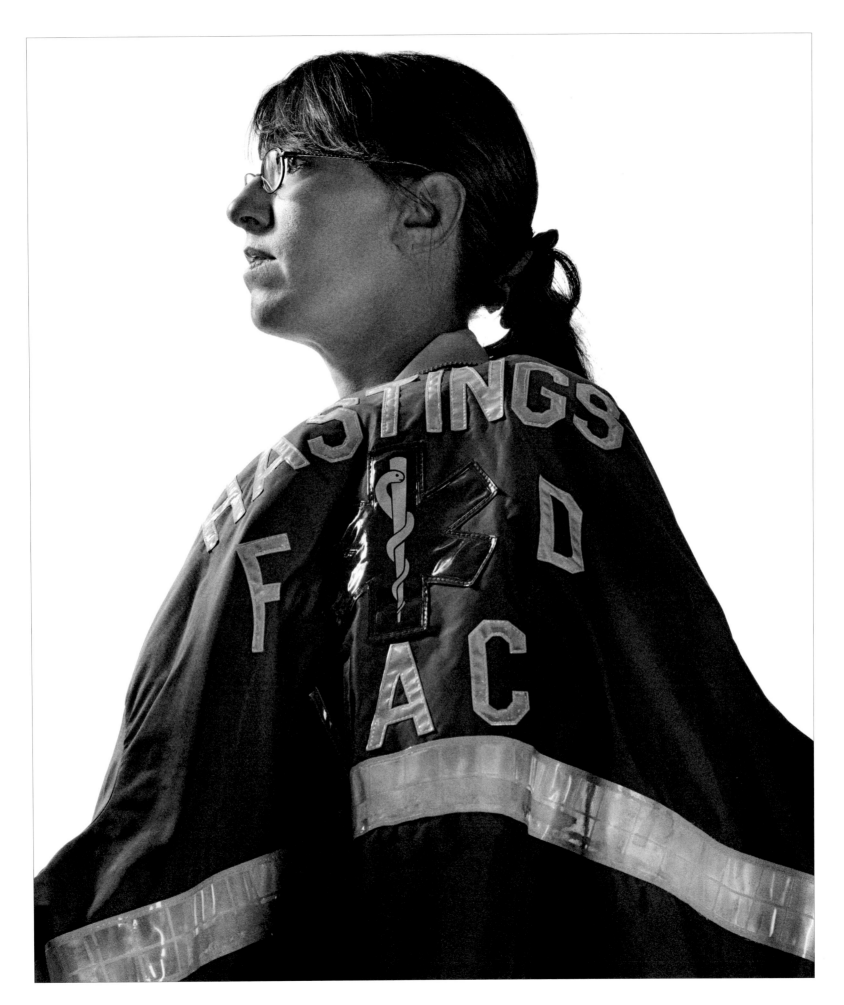

Michele Porter

Thomas H. Drake

TOM DRAKE WAS A BOY when his father died. His mother, a local schoolteacher, did the only thing she could manage at the time — she asked a few of the men on her street to keep an eye on the young boy. Charles "Jerry" Saeger, who was married but had no children, took a special interest. He was the captain of the fire department and took young Tommy to the firehouse often, letting him climb aboard the engine. Other times, he kept him in line. Tommy turned out fine, going off to join the U.S. Navy and becoming a decorated police sergeant and Westchester County detective. Tommy returned to live in Hastings, where he eventually captained the fire department. Saeger was well into his 80s when he died, but had lost his wife, Alba, two decades earlier. At the end, he had no family, save for Tom Drake, who, now a grown man, sat vigil by the old man's deathbed.

SCOTT GOBER, 44, joined the fire department three years ago. His first fire was in the basement of a restaurant — clogged, smoky, and dark. He recalls how his eyes, darting left to right, could not provide perspective; he had no idea precisely where he stood. However, intuition took over.

"I hit the second set of stairs going down and had no clue about how many more there were or how far down the fire was," said Gober, adding that the shouted commands coming at him from all sides gave him a sense of mayhem. "Smoke was everywhere and I was loaded up with equipment, unable to see. I had wanted to give back to the community, but as I was descending into that sinkhole I said, 'Well... Now I'm a fireman.'"

When Gober reached the bottom basement floor, where the fire raged, he was commanded to use his infrared camera to check for fire extension in the walls of a nearby stockroom. Smoke had seeped underneath the walls, but there was no apparent fire within the walls — and Gober took a moment in the eerie stillness.

"It was like I had been airlifted into scene of mayhem, then all of the sudden, it was as if I were underwater. No barking orders, no one was there. In a split second, you can go from one environment to the next. You have to be prepared, but you can't really be prepared."

Scott Gober

John Petas

IT IS NOT FAILING COURAGE, just aging knees, hips, or eyes that causes a volunteer firefighter to retire from active duty. That does not mean, though, that the old salts also give up the essential work of running the firehouse. Well into his retirement years, John Petas, who recently had a heart attack, is the firehouse treasurer. This may appear to be a reduced task, but is no less meaningful than dousing flames, as Petas, who strictly-speaking favors "the less spending is more" ethos of finance, arches a brow conspicuously whenever requests for funds are made. Firehouses do not usually have a lot of money; saving every last thin dime is essential — and an important role for a firefighter who has answered his last active call.

JAMES DRUMM started his adult life as a Marine and will end it as a firefighter. In between, came a lifetime in corporate America. However, Drumm says that you will be remembered by your children not for what you do for work, but for what you choose to do outside of it. At first glance, Drumm has a solemn and cross demeanor, but he is also much more than his military bearing, which is partly theatrics, permits people to see. He's intelligent and worldly and a conversation with Drumm can veer in any direction, from the shores of Tripoli to a recent report on National Public Radio. Politics rarely enters into conversation at the firehouse, but the firefighters make an exception in Drumm's case. He doesn't suffer Democrats lightly and it's a favorite bit of banter among the firefighters to link Drumm romantically to actress Jane Fonda. Another joke around the firehouse is that Drumm joined the department right after the earth cooled, suppressing those initial hot balls of fire. Since then, he has served as firefighter, captain, deputy fire chief, fire chief, and fire inspector. Nearly 80, Drumm, forever a firefighter, still makes nearly every call.

Deputy Chief JT Drumm

SUNSHINE HOUSE

SUNSHINE, COLORADO

VOLUNTEER FIRE DEPARTMENTS ALWAYS train for the big one, sometimes for decades. For the Sunshine Fire Department in the rugged foothills northwest of Boulder, the big one came — and went horribly wrong.

Sunshine, Colorado, is a small district of only a few hundred year-round residents who tend to be enamored of the strong sun and able to bear the abundant snow and occasionally errant winds.

The enormity of the surrounding terrain is no abstraction; to your left, right, and rear are steep untrammeled mountains, but the flatlands, or Great Plains, stretch for miles in front of you. This meeting of plains and mountains does strange things to storm systems, which rumble the clear path along the flatlands for miles and then, hitting terminal velocity, slam the mountains head-on. The collision results in lightning storms with little parallel.

When the lightning bolts hit dry, drought-starved brush, what occurs? Fire. When there are multiple strikes or the winds do mischief and the fire is not contained immediately, what occurs? Trouble, with a capital "T".

Trouble — stark and hell bent — started the workweek of September 6, 2010, at 10 a.m., probably in the most unlikely manner.

THE SUNSHINE FIRE DEPARTMENT got its start in a common way. A band of residents had its fill of slow response times from the neighboring communities who lent a hand in responding to Sunshine's emergencies, but considering the distances, were a long-time in coming. Creating their own fire department, taking matters into their own hands, would solve that. When, for example, John Tveitaraas' chimney caught fire one night in the late 1960s, the wait for firefighters over the mountainous terrain would have been nearly an hour. By that time, his house would have been little more than charred wood and ash, so Tveitaraas and a number of resourceful friends climbed to the roof and shoveled snow down the chimney. The fire was squelched.

Soon after, in 1969, there was an old-fashioned community "barn raising" to build Sunshine's first firehouse. In a gesture of symbolic importance, Tveitaraas and his friends shingled the roof. Though a small department from the start, with rarely more than a dozen active members, the Sunshine Fire Department immediately began responding to all manner of emergency: from car accidents to wild land fires. The firehouse, like most, became a center for local social life. It was as if it had always been there, making what happened incredibly ironic and sad beyond measure….

THE FOURMILE CANYON FIRE, its destructiveness tearing at the community, was started not by a bolt of lightning, as is often the case for wild land fires, but inadvertently — by a volunteer firefighter. Criminal charges were never filed for the accident, which occurred when a 71-year-old man, a member of the fire department for two decades, doused a fire in a pit on his property with water, which he stirred about to make certain no embers were left hot. A few days later, though, slapping winds reignited a few of the embers and blew them about, igniting the fire in the canyon. Fourmile Canyon was once host of a railroad that ran to long-abandoned mines. It was bone-dry.

Once a wild land fire gets going, you pray for cool air and rain and fear erratic winds like the devil. In this case, prayers were not answered. The fire started at the base of Emerson Gulch, but spread nearly unbroken.

Ultimately, the fire burned more than 6,000 acres. Fighting it involved 1,000 firefighters who were called in from great distances to assist Sunshine's courageous but small crew, arriving in airplanes and more than one hundred fire engines. Fifty-foot flames, which firefighters later described as tidal waves of fire, defined the Fourmile Canyon fire. Though no firefighters died, these volunteers were in extreme peril at any given time. They fought through heat and wind and abject exhaustion, leaving their families and careers behind to get a handle on the fire, the likes of which no one had ever seen before. The fire covered over ten square miles, burned for eleven days, and caused the evacuation of thousands. Even a section of Boulder, Colorado, was put on standby for evacuations. In the end, the fire burned down 169 structures, mostly homes. Just under half the houses in the district were in ruins and uninhabitable. A good number of Sunshine firefighters watched their homes go up in flames, without leaving their posts to try to save their own homes and possessions. In the end, damages were pegged at well over $200 million, the most expensive wildfire in state history.

A year later, the ghosts of the fire remain in the vast hillsides, still blackened and charred.

Still present in the hillsides, too, is regret, along with recrimination and accusations that the fire could have been fought more effectively. Even though fighting fires of this magnitude is more art than science and involves variables that have vexed firefighters for far too long, it didn't matter. From the genesis of the fire to its duration, Sunshine's population, many without homes, cast blame.

"It was clearly an accident, but especially as a volunteer firefighter that's a huge no-no," said a resident, about the start of the fire. "We don't even burn tiki-torches up here. You think you would know." Her comments, given publicly in the *Denver Post*, were one of the more polite assessments heard in these parts.

Other stories tumbled in, mostly involving the Sunshine firefighters who had, from a distance, seen fire encroaching on their own homes, but remained at their posts to save other people's homes. Even the fire chief lost his home. There were also those who helped evacuate older residents when they could have been accompanying their own families to safety. Moreover, though the level of property damage was enormous, no one died. Everyone was evacuated to safety, many with the help of Sunshine's small but valiant cavalry, which stood their ground in face of those 50-foot walls of fire. Less than a year later, recrimination was at least tempered by gratitude — well-earned gratitude for a long fight waged against a furious and defiant fire.

IN 1979, THERE were no women in the
Sunshine Fire Department, but Linda
Ballard, now 66, looked out her living room
window one clear night only to see a 50-foot
pillar of fire piercing the sky. "Nothing in
my life had ever been that gripping, thank
God," she said. When she learned that all
of the firefighters fighting the blaze were
volunteers, she knew she had to join. Being
a woman firefighter in those days was not
without its difficulties. The department was
a bit of a Boy's Club and there were always
those assumptions — sometimes subtle,
more often overt — that she couldn't keep
up. Ballard can't even remember how long it
took for the second woman to join, but she
forged on — happily — in the service of the
community, with the memory of that pillar
of fire seared in her mind. After a decade, the
sight of a woman on the force was at least
no rarity. Linda's family was not surprised
by her volunteerism or her adventurousness.
She had previously worked avalanche rescue,
as well as the Civil Air and Ski Patrol.
Later, her husband Henry followed her into
firefighting. For years, the couple juggled
responding to fires and childcare.

DRIFTING SIDEWAYS in retirement would have
flustered the 56-year-old Henry Ballard. So
instead of answering the limited number of
fire department dispatches that he did when
he was commuting to Denver five days a
week, working as a software programmer,
he took his retirement package and threw
himself completely into firefighting. Ballard
joined in 1986, seven years after his wife,
Linda — "Lin," to him — became the first
woman on the force. The couple never
fought fires shoulder-to-shoulder, as one
would always have to stay back to take care
of the kids. Henry's mother, Lee Ballard,
is 87 and still lives in Western Montana,
outside Missoula, where he was raised. She
didn't raise her son to be a firefighter, always
wanting him to do indoor work, where he'd
be safe. As a boy, Henry didn't even take
many risks, but as a retiree, she can't keep
him away. In the summer of 2011, Ballard
volunteered to assist in fighting the deadly
wildfires in far-off Arizona, which hit the Far
West like a pipe wrench. His wife, a fellow
firefighter, gave it her blessing. On the phone
with his mother just beforehand, though, the
toll and scope of the Arizona fires came up.
"You wouldn't want to go there," she said,
sensing her retired son's temptation.

Linda Gail Ballard & Henry William Ballard

Michael Schmitt

MICHAEL SCHMITT, an immigrant from northwest Germany, near the Dutch border, volunteered to fight fires partially as a means of giving back to his adopted community and country. The stint was interrupted, though, when a divorce, and ensuing immigration paperwork issues, landed him back in Germany. As soon as he returned to the U.S. four years later, however, he signed on again.

A self-professed adrenaline junkie, Schmitt always got goose bumps at the possibility of experiencing a giant wildfire...that is, until he saw the Wonderland Lake fire in July of 2002, which endangered everything from firefighters to waterfowl. The giant flames reflected a devilish red set-off against the lake and night sky, giving the entire tableaux the mix of flaming death and colorful beauty. The fury and force of the fire was no abstraction. Neither was the speed. Schmitt's courage did not fail in the face of the wildfire, but if he never sees another roaring blaze that kicks off his adrenaline rush for the rest of his life, he won't feel cheated.

WOLVES ARE NEVER at the door when you are a pharmacist. Mike Sanders has never gone without work — after all, dispensing pills and ointments is as cut and dry as professions tend to go. "But lightning strike fires, now that's a crazy business," said Sanders, remembering the hot, dry day a few years back at the Bender's property, when a thunderstorm rolled through and they almost lost the handle on what came next.

The Old Stage fire, too, in 2009, which burned 1,000 acres, forcing the evacuation of 500 homes, had the devil in her. In that one, Sanders was in a strike-force of five engines headed up the mountain. There were flames all along one side of the road. He was in the second to last engine of a slow ascent, and Sanders felt trapped as the wall of fire surrounded him. He was certain that the engine would become his tomb. Through sheer luck, the fire stayed at bay.

At least when faced with a 50-foot wall of fire — as like in the Fourmile Canyon fire — the pharmacist, who gets punchy when recounting such predicaments, knew that if the situation turned even less predictable, he could take refuge in a nearby pond...the only place where the fire could not touch him.

Michael Sanders

Peter Beresford & Rick Lansky

A SPINDLY INTERNET application programmer with no aspirations as a child or an adult to be a firefighter, 44-year-old Rick Lansky's introduction to the inside of the firehouse came about when his wife volunteered him to build the department's website. As soon as he entered, he was taken — a shift he still can't see with any clarity. "It's like your driving along the road," he said, "when you get close to the edge, it just sucks the car down." The website could wait: Lanksy donned bunker gear. Several years later, he received a call for a structure blaze. No details, not even a house number. He rushed up South Canyon Drive to the sight of 20-foot flames shooting out of the roof of the fire station. In the ultimate irony, an old faulty water heater had ignited. The flames spread to the back of the station and before long had blown out the back roof, spelling a total loss for the structure.

"You are supposed to go the station to get the engine," said Lansky, "but what happens when the station is on fire?" Improvising, Lansky sprinted in and saved the engine.

"I remember a lot of moments from that fire," he said. "None of them good."

STEVE WALTMAN, a 46-year-old physicist whose first job was in the Time and Frequency division of the National Bureau of Statistics (where he measured the color of light to considerable levels of precision), knew he was going to have to fight three men. Stationed alongside a road deployed as a barrier against the massive Fourmile Canyon fire's spread, Waltman realized that the scale and scope of the fire that surrounded him were "more than I ever wanted to deal with," he said. He tried to envision a good resolution: "If I could not imagine it, how could I hope for it?" The winds were hitting hard, though, and then he caught sight of it in the far distance: the fire creeping up the steep slope below his house. He knew he had enough time and resources to save his house, or at least essential possessions. It would, though, entail leaving his post — and that fistfight. More than that, though, was his sense of valor and responsibility to a job that didn't even pay him a cent. "I was true to my training and stayed," he said.

Former Chief Steve Waltman

Donald E. Dick

Sheret Matheson

Caleb Sevian

AT 41, CALEB SEVIAN is considered "one of the kids of the department." A money manager by trade and a weightlifter and endurance athlete by choice, Sevian's relative youth, strength, and fitness did not, however, put him in any better stead than his colleagues during the Fourmile Canyon fire. In his mind, Caleb still sees the harrowing incident as an "ember storm," as if a sky of stars had turned fiery and angry and then fell. Fighting the blaze, he came up a gulley to the sight of three houses on fire.

"All three were fully engulfed," he said. "I was in a state of stock. I had entered a surreal world, and once I entered, I didn't know if I could leave it again."

He did. No firefighters were killed in the blaze, but those three houses — and dozens of others — were gutted and lost. The thought of that still haunts Sevian.

"You join the fire department to do good, but I feel embarrassed that in our district one-third of the houses burned down," he said, waving a hand and looking toward a hillside, still scorched nearly a year later.

Perhaps thinking of the homes that made it, or merely reflecting the firefighter or endurance athlete's need to see the glass half-full, Sevian added: "We did some good that night too."

GEORGE WOODWARD'S FATHER, also named George, was a career firefighter in St. Paul, Minnesota, where, in the frigid winter air, he barreled into burning homes and up the narrow stairwells of wood tenements. The elder Woodward was a man of few words, but would always tell his son, "You gotta' do what you gotta' do."

The son, now 53, had no interest in making a career out the fire service — office work was safer and saner — and instead went into telecommunications management. However, when his own children were grown, the younger Woodward finally followed his father's path, volunteering to fight fires in the Colorado mountains. Then his son, Kyle, in his early 20s, joined.

Woodward was driving home from a Utah camping trip when the big Fourmile Canyon fire hit — and Kyle called to tell him that the fire was massive and expected to last seven days. "I told him don't become a memorial marker," Woodward said, "but like his grandfather, Kyle is a man of few words. All he told me is, 'I'm headed in.'"

George Woodward

Bruce D. Honeyman

A PROFESSOR EMERITUS in environmental engineering at the Colorado School of Mining, Bruce Honeyman has served as both President of the Faculty Senate and Chief of the Sunshine Fire Department. To Honeyman, 57, the jobs could not be more dissimilar. "Chief," he said, "is like being mayor. Someone calls you, 'the dog is sick, the horse is out of the barn, the neighbor is shooting a gun.'" And the faculty Senate? "Well, when there is an emergency in college," Honeyman said, "you form a committee. In the fire service, you act."

On occasion, however, even when responding to emergencies, there is nothing that can be done. Honeyman dissolved into tears at the recollection of responding to a call in which a teenage boy with muscular dystrophy, already dead, was laying in his bed amid toys. The boy appeared melted into his sheets.

Honeyman was also chief during the Fourmile Canyon fire. He lost his home in the blaze, but the Faculty Senate President turned Fire Chief fought through the night, amid grief and exhaustion, to save the homes and lives of his neighbors.

AS A BOY, Chief Brett Haberstick, a psychiatric researcher on the genetic component of addiction, wanted to be a cook, but eventually decided against it. He had an inordinate fear of getting burnt. Clearly, he's gotten over that: Today Haberstick is chief of the Sunshine Fire Department.

Although volunteer firefighters want to perform good deeds, noted Haberstick, there is also some facets of an addictive personality in the job. "There is a certain amount of novelty and sensation seeking," he said. "Obviously, when things go bad in firefighting, they go really bad."

That dark possibility can make a firefighter feel sharp and keen, at least in the moment. Emotions that languish in the vast majority of life, so moored to normalcy and small tasks, can soar during a fire. For Haberstick, the physical aspects of firefighting are a welcome departure to his bookish, research-centered vocation. In fact, he feels it even makes him more competent at his day job.

"I perform better research when I've been active," he said. "Of course, I could jog on a treadmill, but here you also help people in their worst moment."

Chief Brett Haberstick

Steve Stratton

SEA RANCH HOUSE

SEA RANCH, CALIFORNIA

PERCHED ALONG A RARIFIED ten-mile stretch of cliffs overlooking the Pacific, Sea Ranch is a remarkably well-to-do community where, at least economically, little ever breaks bad.

There is, though, uncertainty — a collective concern, if you will — about the cliffs. Though they deliver majestic ocean views, they also stand as ruthless hazards, especially during the summer when vacationers, unaccustomed to the sudden drops, arrive. Occasionally a visitor, most often at night, will fall down to the wave-beaten rocks, as if stepping onto a trapdoor. They have to be rescued or, in the worst case scenario, their bodies retrieved.

However, it's not just the perilous nature of the cliffs that gives cause for alarm: During times of limited rainfall, wildfires are also a factor — and even bastions of such privilege, where lives are well-lived, have house fires. Fires in off-season vacation homes are often not reported until the fires are fully involved.

Sea Ranch faces all of this, but also has an additional challenge: It is made up of just over 1,000 full-time residents. It is a small area, obviously an insufficient population to sustain a paid fire force, but also nearly too tiny to gather up a sufficient number of volunteers. In fact, there are only nine members on the force.

Moreover, successful people tend to come to Sea Ranch to retire. This means that a large fraction of the population is older, with the average age tipping past 55. Since volunteer firefighters are culled from the general population, those who battle blazes and repel down cliffs to rescue those who tumbled — or swimmers caught in riptides — are older too. In fact, most of the firefighters at Sea Ranch are in their 60s and 70s. The youngest is 54. You won't see inky hair, thickset biceps, or youthful swagger in this firehouse.

"There aren't many of them" said a resident, "and those there all seem — as the rest of us — like they should be in assisted living. But they get the job done. Day in day out, year after year, they get the job done."

In many ways, the Sea Ranch Fire Department is typical. They run fund-raisers, holding community barbecues, letter-writing campaigns, Christmas tree sales, and craft fairs, as well as selling Sure Fire Recipes, their own cookbook. There was even a cabaret evening held at the local community center. Save for the redwood siding, the firehouse is typical too: A garage bay with room for four rigs, a classroom, an office, and a small kitchen.

On the other hand, the fire department appears to put on a daring display of rank lunacy. They are a bucket list gang speeding in fire engines, rushing into burning buildings and repelling down wave-washed rocks, under rising tides. Several of the firefighters are retired powerful corporate officers, though they are led by a retired high school English teacher. However, together, in this self-styled and privileged community three hours north of San Francisco, these firefighters along the Pacific cliffs protect their remote outpost with aplomb, born from ageless dedication.

A RETIRED HIGH SCHOOL English teacher who rose to chief after six years on the force, Michael Scott knows that it is possible, if not probable, that every fire call will at some point hit a snag. Firefighting simply has more variables than teaching Chaucer to teenagers. A home on fire can collapse suddenly — it's impossible to gauge its structural integrity from afar — or there can be a heartbreaking ending. Such was the case during abalone season a couple of years ago.

In this annual ritual, an influx of sport fishermen, donning wet suits and diving for the delicacy off the coast of Sea Ranch, fill the local homes offered for rent. At night, there came a call for a missing person: a mother of two who was visiting with her fisherman husband. It was foggy and damp and Chief Scott knew the dangers that an errant back-step posed to outsiders not used to the area's treacherous cliffs. He pointed toward the bottom of a nearby bluff. "I told everyone to look for bright clothing down there," Scott said. They found her between the rocks and surf. She had merely gone out for a smoke. They brought her body up in a Stokes basket; Scott had to tell the horrible news to her family.

"(It was) one of the first times I ever dealt with death," said the retired English teacher. "Since then, there's been a few more."

Chief Michael Scott & Max

Michael Tuft

A FIGHTER PILOT in the Navy during Vietnam, Michael Tuft hurtled off the sliver-thin decks of aircraft carriers, a task he thought gave him enough of an endorphin rush to last a lifetime. After the war, he became a corporate counsel for a bank, where the lack of action tinged with risk made him feel a bit drawn. When he turned 65, he retired, moving from Sacramento to Sea Ranch, where Tuft became a firefighter. If he was looking for an occasional taste of danger, he found it in this job: One night, a young mother of two, out for a nighttime smoke in the meadow across from his home, lost her footing and fell down a cliff to her death. Tuft rappelled the high walls to retrieve the body.

However, most of the time, firefighting is not as deadly as war. Yet, there was that time when he was riding in a tight convoy of fire engines, advancing up a hill during a wildfire, taking on a blaze that was being propelled forward by the night winds and creating its own red-hot gusts of air in response. Between the fire to the right and the mountain on the left, Tuft's rig was nearly pinned. "There was no way to turn the engine around," he said. "It was like Dante's Inferno and I thought we were dead." Luckily, the fire did not advance past the road and Tuft's convoy returned safely. He's barely missed a call since.

BILL LAWSON, 78, was an ingénue of 67 when he joined the Sea Ranch Fire Department. His family greeted the news "quietly," he remembers. "No one really said anything. Even though they have been extremely supportive since, nobody really said, 'Yeah, great, go ahead!'" With hackles at home only slightly raised, Lawson became certified.

Calm at a fire scene, he says, is paramount, though it is also an ongoing struggle to achieve. Sometimes — even decades removed from childhood — the excitement of the flashing lights and sirens means all Lawson can muster on scene is a calm façade while his stomach churns and heart beats faster. However, he knows that the calm, even if forced, can help the victims who are looking for any sense of assurance that all is well and they will emerge from the chaos.

Lawson was born on Olympic Boulevard in Los Angeles, cattycorner to a firehouse, where he was once allowed to haul the hose after fruit boxes burned in an abandoned lot. He was impressed by the steadfast dedication of the night captain. Even off-duty and asleep during the day, the night captain, who lived about a half-block away, would burst from his bed and show up at the firehouse when the engine rolled. He was always ready to jump aboard. In recent months, Lawson, a retired high school history teacher, has suffered from arthritic stiffness and is having some problems moving about. Unable to go out on fire calls, he's taken to sweeping the fire station. That old night captain, Lawson knows, would be proud of his commitment.

Bill Lawson

Walt Jorgenson

WALTER JORGENSON, who went from serving in Vietnam to laying phone lines in Haight-Ashbury, is 68 and just recently got flame tattoos affixed to his forearms. If war and even a severe case of culture shock couldn't kill Jorgenson, he thought at the time, perhaps nothing ever could — and he'd always be young. "When I turned 65, I said 'I'll never grow up. And I'm on the fire department. Then I figured — well, name one good reason why I don't have flames on my arms?'" Jorgenson, who exudes suspended adolescence, still gets hopped-up when he sees the flames on his arms, though explodes into scratchy laughter thinking about how his 68-year-old skin sags and makes the fire appear to be going out — fitting for a firefighter.

With well-honed stories about everything from laying telephones lines around piles of sleeping hippies to his current incarnation as a real estate agent, Jorgenson's favorite yarns are about the fires that almost got away. There was the one two decades ago that started when a resident discarded spent barbecue coals into tall grass. "The fire jumped the highway," he said, "right toward a batch of homes." That was before the prevailing northerly wind shifted. "That was hairy," he summed up, before considering the many years ahead and adding: "One of the hairiest yet."

FOR 60-YEAR-OLD Randy Burke, the notion of serving his community by sitting on another steering committee was as appetizing as week-old ground chuck, so five years ago, ready for something more adventurous and exciting, he joined the fire department. His grown children chided him for this somewhat rash move, unusual for Burke, who still works running the local water and sewer lines. His son told him he was crazy and should be off fishing. His daughter told him he was crazy too — "nuts" was the precise word — but allowed that the thought of her old man as a firefighter was cool.

The first time Burke really had to buckle down as a firefighter was the Timber Cove fire two years ago. Electrical lines fell on dry shrubs along a bluff and the fire, harnessed to the wind, sounded like a run-a-way freight train; the blaze eventually extended to the tops of trees. On various hose lines, Burke sprayed water this way and that, up and down, for hours. To this day, he has a blinkered perspective on the time during the fire. "It wasn't dangerous for all nine hours," Burke said, "and it felt like it lasted just a half an hour. But I know this: afterward, it took four or five days to calm down."

Randy Burke

Bonnie Plakos

AT 54, BONNIE PLAKOS is the baby of the bunch, the youngest firefighter in the department. She lived most of her adulthood in Los Angeles, where to better navigate gridlock in the land of freeways, she drove cars the size of tuning forks. Now she drives fire engines. As a former software architect whose workplace was generally cubicles, the movement, speed, and sense of raw power on the big trucks in rural Sea Ranch are reparative therapy for the longtime small car, city dweller — even intoxicating, though with a caveat. There is often not enough space. There are small dirt roads in Sea Ranch and sharp curves perched over cliffs. A careless turn can turn deadly in a hurry. To master driving beasts above the abyss, Plakos takes the fire engines out for hair-bending test runs. She drives up in the hills and she drives backwards. She checks out the desolate little gravel roads and takes side trips to scout out particularly remote homes or gauge potholes, as well as notoriously slick curves. "Everyone thinks, we're a bunch of old guys — what can we do? But we can drive equipment, pull hose, and do cliff rescues," she says, before hopping up into her rig.

WHEN THE ACCOMPLISHED founder and CEO of Advanced BioNutrition Corp. retired from Baltimore, Maryland, to Sea Ranch and the fire department, David Kyle went from chief decision-maker to mere cog. He hasn't regretted it.

Kyle was in his late 50s when he joined the department and became "the new kid on the block," as the former biotechnology industry rainmaker put it, a rookie marching along the bottom of the learning curve, reporting for duty to wait for commands. Taking orders was a departure for Kyle, now 58, though a surprisingly welcome one. Decisions at work — to hire and sometimes fire — had, truth be told, always weighed heavily on him. Besides, work at fire and accident scenes was a team effort, which came harnessed to a heartfelt camaraderie that was largely absent in the cutthroat world of business. "It was just a high-pressured life style," he said, shaking his head nearly a year after leaving corporate America. Claims that companies are families don't always hold up to harsh light, he added.

Still, Kyle never thought he'd be a firefighter, certainly not as a boy growing up along the prairies of Saskatchewan. However, he got the bug many years ago when he was in a car crash and had to be extracted from the vehicle by a team of firefighters who, in effect, saved his life. "You see it in the movies and all the time on television, but at all these accident scenes and fires people put themselves in jeopardy to help out," says Kyle. "That was a bit of a discovery for me."

David Kyle

David Rice

IT'S A DELIBERATE UNDERSTATEMENT to call Sea Ranch isolated. With the nearest hospitals hours away, remoteness in this little community of 1,300 is akin to an unseen enemy. David Rice, 65, a retired CFO of several software companies, got an unexpected perspective on the situation a few years back, when he fell, dislocating his elbow. His wife called 911 and, despite the seclusion of their home, a dozen firefighters and EMTs had seemingly materialized in their living room within minutes. Rice was in pain, but markedly impressed by their dedication and competency. The challenge was not put to bed, though, with their appearance. There was still the matter of "busting their buns," as Rice put it, to get him to a hospital in Santa Rosa, a good 80 miles away. For Rice, only an elbow was at stake. The experience, though, was a revelation: lives depended upon such responses. These days Rice, working as a firefighter and EMT, does the materializing in living rooms, helpings others make the long trip to the hospital, as he did in the middle of the night once when an older woman was bleeding from the head. Then he came back to Sea Ranch, where a working fire was in progress. But the former Palo Alto resident refuses to slow down despite the long nights he sometimes endures. He recently completed a course that required him to rappel down a seven-story tower, a helpful skill to assist in cliff rescues in this distant outpost.

THE 63-YEAR-OLD decided to join the fire department two years ago, but figured it would be the same claptrap. Richard Pfeifer, a mild-mannered soul who spent a career in telecommunications finance and regulation, would, he thought, lend a hand by serving on a department committee. But once on deck, he couldn't resist the boyish pull. He learned how to fold hose and pump water and grew awed by the physics of fire. "Every time I go on a call," he said, "I learn something new." Before too long, he was working the pump panel, controlling the water flow to various hoses to douse a blazing house fire. "In life," he said, "You do things you never thought you'd do." Pfeifer still sits on the Sea Ranch Chapel board, which serves to maintain a local chapel in good stead. He is also a member of the Sea Ranch Investment Club. But, said Pfeifer, throwing an eye out toward the rigs, tools, winding hose lines and fellow firefighters, "this is my real passion. Now I want to do this for as long as I can walk."

Richard Pfeifer

SPARTA HOUSE

SPARTA, GEORGIA

NEARLY A HALF A CENTURY after the desegregation of Major League baseball, the United States Military, and southern schools, the Sparta Fire Department, in a county toward the eastern end of Georgia that is nearly all black, was exclusively white. Today, only 15 years later, the Sparta Fire Department's assistant chief is black.

Around the firehouse today, the assistant chief is commonly seen laughing it up with the current and former chiefs, who are both white. The three count each other as best friends. Members, both black and white, who work side-by-side on hose lines, speak of maturity and fostered understanding and evolved, if not overdue, notions of fairness.

"We've all matured a heap over the years," said one firefighter.

The story of race in America, even in altruistic endeavors like volunteer firefighting, is never simple and linear. Progress, while undeniable, has inevitably pulled up short of total. Perhaps bruised histories never quite heal or just maybe, given a bit more time, everything will be just fine.

SPARTA SITS IN HANCOCK County, the poorest in the state. The median income for a household is only a hair above $20,000 and well over one-third of all adults never graduated high school. In Sparta, the county seat, nearly half of all children live below the poverty line. For firefighters, poverty means limited equipment. Paired with the presence of inexpensively made trailer homes, many down isolated rural roads, and there are further challenges for firefighters.

The district covers a cavernous 475 square miles, with long distances to travel leading to low response times. Due to the lack of fire hydrants in all but its most concentrated locations, firefighters are forced to draft water from ponds and streams, shuttling vehicles between the fire scene and nearest body of water. To augment their manpower, the Sparta Fire Department uses prisoners from Hancock State Prison. However, even that only does so much good. Though the prison is right around the corner, with security checks, the auxiliary force of convicts can never respond quickly enough. Over the decades, the district has seen more fire-related deaths of residents than most volunteer departments.

Challenges abound, but it's not just luckless circumstances at play. Some of the department's challenges are self-inflicted. For too many decades to count, the South's old ways were durable at the Sparta firehouse, which was insular and stand-offish, at best. Operating as an all-white fire department in an overwhelmingly black community — traditionally one of the blackest counties in the nation — led to many problems, including chronic mistrust between the firemen and members of the community they had volunteered to serve.

Calls to, say, housing projects populated by all blacks, many felt, were not answered as persistently or in as timely a manner. Nearly irretrievable mistrust developed between the black community and the white fire department. At fire scenes, the firefighters, under suspicion for taking their time, were met with eruptions of anger. Residents surrounded their fire engines, shouting accusations.

"It was pandemonium at times," said one firefighter. "Talk about raw nerves — everyone was flustered. I remember trying to put on a fitting for a nozzle, but my hands were shaking from the shouting and I could barely hold it."

It was a morass, with each side deeply alien to the other. That would begin to change in the 1990s.

One day in the mid-1990s, all those years after integration of other institutions, a black man in his mid-20s named Mario Chapple walked into the firehouse and asked to be a member. Much to the public's surprise, Chapple was accepted. There were no legal tussles or public protests involved — only a solitary black volunteer firefighter, who went to work answering more calls than most. Today he is an assistant chief.

Such a turnaround would normally merit exhilaration, but there are no neat and tidy storylines at play here.

Whether to iron out underlying tension or just present a good front, no firefighters speak of any lingering ill-will between the races. Yet since Assistant Chief Chapple became a member — and even with him in charge, though not, as yet, chief — there are only a small number of blacks in the department. Chapple acknowledges that even today, black friends need assurances that they will be welcomed.

Perhaps it is true that a bruised history will never completely heal, but every time another black is recruited or Chief Keith Webster, who is white, interviews a white recruit and tells them to check any unwholesome stances on race at the door, the Sparta Fire Department, which faces so many practical challenges operating under such difficult circumstances, is pushing their self-inflicted trouble further into the past. Progress continues, as best it can.

KEITH WEBSTER, the current chief and a local pharmacist, remembers the bad old days when the all-white department had to respond to fire calls in the black housing projects. Although it was never confirmed, many of the people living in those apartments harbored suspicions that the volunteer firemen responded more slowly to black parts of town and that the firefighters were less willing — even unwilling — to take risks. At times, crowds would form around the fire engines, shouting insults.

Some in the black neighborhoods still feel this way, but Webster says that it helps that Assistant Chief Mario Chapple is black and goes out to the community to address any concerns about response times. Webster, who is best friends with Assistant Chief Chapple, has a message to recruits: "If you have any bias, check it now."

Chief Keith Webster

Devoris Lamar

IN A CATASTROPHIC TURN, Devoris Lamar's family lost their home and all their possessions in a fire. Their life became a daily nightmare of holding up at friends' homes and financial ruin. Depression and self-pity could have simply metastasized, and Lamar, who was in his early 20s, could have chosen to simply move on, remaking himself someplace else with better job prospects, hoping to forget about the fire. Instead, he called the firefighters he had seen rushing into his home in a vain but valiant attempt to save it — and he joined them.

Today, a few years later, his family is still struggling to find their footing. They had no home insurance and currently have no home and very few possessions. Lamar was even laid off from his job twisting yarn for a carpet manufacturer. Yet, there he is, rushing into the fiery homes of total strangers, seeing in his own tragedy a call to service.

"Mom and dad," he said of his parents, "are pretty proud of me."

THOUGH JASON BUTTS, 39, has only been
a volunteer firefighter for two decades, he
spent what he calls "a lifetime" around the
firehouse, thanks to a family legacy, even
playing on the trucks as a child. That was
way before the Sparta Fire Department,
in an overwhelmingly black county, was
integrated. Back then, imagining blacks and
whites working alongside each other was a
laughable proposition, Butts said, adding
that keeping races separate was the only
way to keep peace in the valley, as they say.

What shocks Butts the most about
the integration of the department, which
now has a black assistant chief, is how
easily it all occurred. In his eyes, it has not
been a big deal.

"We all have to keep our wits about us,"
Butts said with a shrug. "The call comes
and we're all just together."

At that Butts pauses before continuing:
"It's just all come a long way...We've all
matured a heap over the years."

Jason Butts

Camaron Brown

UNEMPLOYMENT MAKES FOR a life of protracted frustration, but firefighting serves as a cure, at least for Camaron Brown, 21. With the economy fractured, Brown picks up a bit of handy-work here and there, but barely even makes a subsistence level wage. This gives him too much time, he admits, to lie on the couch and watch TV. That is, until he has to rush to answer a fire call. "I just like the trucks and the excitement," he said. What's more, he added, "I'm doing something instead of nothing. Like it? I love it — and need it."

Al Butts

Mickey May

IT WAS NEARLY A decade ago, but 40-year-old Jim Brake still bears the psychological scars of the apartment fire on Dixie Street a week before Christmas. One night, a mom went off to the store, just for a moment, leaving her two sleeping children behind. The fire hit and the staircase to the apartment was burned, leaving no access from below. However, there was access from above, because the roof, too, had been destroyed by the flames. Brake was suspended in a ladder above the roofless building. That's when he found himself staring directly down at the children's charred bodies. They almost looked like they were sleeping. That image haunts Brake, especially around Christmas. "Some of us have had a tough time being with that one," he said. "I won't ever escape it."

Jim Brake

Travis Wells

WANTING TO DO GOOD for a county that has so little, Travis Wells, 39, joined the fire department a dozen years ago. He runs a logging company just out of town and rushes back when he can for calls. Then there are the ones that come in the middle of the night, when he leaves his wife and four children behind in bed. Leaving his family — instead of work — fills him with dread, as does the night itself, which tends to conceal more at fire scenes. To Wells, nighttime fire calls pulsate with risk. On weekends, during the day, he brings his children to climb aboard the engines and wonders how he'll react if they ever say they too want to do something good for a county that has so little by serving as firefighters. "I'd be a little worried and a little grateful that they want to volunteer to help the community," Wells said, adding after perhaps considering those nighttime calls, "but more so worried."

JOHN BOYER IS 21, works at a feedlot, and can't easily leave behind the memories of the basement fire that almost took his life. He was the first on the scene — arriving even before the chief — and, though, well trained, forgot everything in the adrenal rush. "I busted the door down," he said, "not knowing what was on the other side. I reacted stupidly," he said.

The door was the only way in. Once he advanced a good way in, he realized that the door, now far behind, was also the only way out. There was not a single window. I thought, "This could be the last place I see."

Boyer had entered a fire alone, without a partner, a cardinal sin of firefighting. Alone, he did manage to find his way back to that door, emerging from the basement and his youthful discretion, vowing to be better tethered to his training and more reverent of danger. He is currently, with the blessing of his family, attempting to become a full-time firefighter.

John Boyer

Justin Smith

JUSTIN SMITH, 23, wanted to become a firefighter thanks to his mother, who was a DeKalb County dispatcher. Though many mothers bite their nails to the quick at the thought of their children becoming firefighters, Smith said his mother was all for it; she felt he'd be less likely to spend his time carousing or drinking. He still manages to fit in his share of partying, said Smith, so his mother's strategy was not a total victory.

However, Smith also can be found doing heroic things, like the night the local flower shop went up in flames. Smith was positioned in an alleyway between the shop and a nearby home. He prevented the spread of flames by soaking the side of the home that abutted the fire. The work, twelve hours worth, nearly trounced Smith.

"It pretty much matched all my expectations of a bad fire," he said. "Bad."

Joe F. McNair, Jr.

Will Webster

Sterling Aycock

FOUR YEARS AGO, Sterling Aycock, 26, was at his mother-in-law's house when he got a call for a fire — at his own home. Rushing there, still a few hundred yards away, he saw smoke rolling up to the sky. When he arrived and was getting into his bunker gear, fire rolled out the front door. "That ain't good," he said, to no one and everyone. He ran around the side of the house, a single-wide trailer, and cut the electricity, but it was too late. His home and everything in it were completely lost. All that he had left were the clothes he was wearing. In a moment, his entire life had unspooled. Aycock couldn't even salvage the photographs on the mantle. Clawing his way back has not been easy. A truck driver, he's currently unemployed and spends the bulk of his day "sitting around with a pager right beside me," he said, hoping to save another family from sharing his troubled fate.

THE TRAILER HOMES that are responsible for a large number of the calls to the Sparta Fire Department are, in terms of construction, somewhat problematic. They aren't built with the workmanship of typical single-family homes. Compromised by fire or heat, they tend to collapse.

Robbie Hudson, 32, was on a hose line when a doublewide trailer did just that. He had taken the line into the trailer because he thought that there were children inside, but there were no children and the empty trailer could not bear the fire and heat. A wall between the kitchen and living room heaved and then crumbled on top of him. Hudson remembered thinking: "I'm in over my head even more than I thought." He was able to budge the ceiling and, although burnt and disoriented, he followed the trail of the hose line back to safety.

Robert T. Hudson

CHIEF
TOM ROBERTS

John (Tom) Roberts

132

TOM ROBERTS HAS been on the force since 1973, including an interlude as chief during that unspeakable year two decades ago: When the district suffered the eight deaths of residents in the first four months of the year. Among them, there were the three children, the oldest only 14, who were trapped in a bedroom, screaming when the fire engine pulled up. The silence that followed the screaming, the sound of that moment of death, still visits Roberts in his dreams. Shortly after that, a burning log rolled out of a fireplace and onto the rug in a home where an older couple was sleeping. The woman nearly made it to the front door before she collapsed. The man was found sitting up in bed. After that was the mobile home fire that took the life of a man and his infant daughter and then the electric heater that put a bedspread on fire, killing another person. Despite these horrible incidents, one after another, Roberts says, he never once wanted to quit. "But I just as soon fire departments didn't have any reason to exist. In firefighting, the law of averages eventually swings against you."

Brannon Brown

James Mashbum

STEVEN ROBERTS JOINED the force in his early 20s, but moved to Macon, some 50 miles away, to become a police officer after getting married. Three years later, his wife died of a brain tumor. She had tightness in her hand and a headache and, in what still sometimes seems to Roberts like a very bad dream divorced from reality, the doctor told them that she had terminal cancer, not carpal tunnel syndrome as they had assumed. After her passing, Roberts moved back to Sparta and rejoined the fire department. "My dad was in the Sparta Fire Department for thirty-three years," Roberts said. "I grew up in it and around it and just had to come back." His wife's death, he said, was too much to handle, "but the fire department is a life sentence of a different kind; it saved me." In those first months back in Sparta, among the few moments that he could escape the sheer pain of the loss of his wife was when he found himself sitting on the bumper of the big fire engine on the apron of the garage, nursing a beer. It was a simple way to pass time — in a life that had become anything but simple.

Steven T. Roberts

Former Chief Robert C. Allison, Jr.

AT 55, ROBERT C. ALLISON, JR. is a changed man, a far cry from his reckless youth. A long time ago, in the 1970s, Allison was a self-professed "hell-raiser who got into messes." In fact, at 19, Allison was in prison, doing 18 months for "never you mind why." Then soon after getting out, he wrecked his motorcycle and barely got away with his life. "I reckin' you could call me a drunk," he said. At 30, though, his son, RC, was born, and Allison straightened out. He worked as a pipe worker, boilermaker and steam fitter and playing ball frequently with RC, who loved the fire house and made pencil drawings of his dad putting out a dragon's fiery breath. Allison has a reputation around the firehouse for fearlessness and a tough gritty veneer, yet can be surprisingly sensitive to other people's feelings. In fact, many credit him with easing the department's racial integration. Asked, Allison, now a retired chief who collects miniature axes, doesn't take any credit. He just shrugs and credits God's will. He has no answers, though, for what happened to RC. One night RC, who lived next door to his father, was partying. Friends called to tell Allison that his son had collapsed. Allison ran over to find his son in cardiac arrest, but his CPR rescue training couldn't revive him. "He was as healthy as a horse and bigger than one," Allison said. "But I couldn't bring him back. I didn't stop working on RC, but just couldn't bring him back."

MARIO CHAPPLE, now 39 and Sparta's assistant chief, was the first black firefighter to join the department. The year was 1996, long after the integration of the military, schools, and even baseball. "It had to change around here," Assistant Chief Chapple said, "but what you have to understand is that everyone — everywhere — is scared of change."

When Chapple first joined, his black friends did not believe his claims that fellow firefighters treated him not only with respect, but also as a brother. In time, though, a few friends agreed to join and found that he was right. The firehouse is still a long way from reflecting the area at-large, which is nearly all black, but these days, Chief Chapple can often be found sharing jokes with Robert C. Allison, Jr., the former chief, each with an arm on the other's shoulder. Less than a generation ago, that would have been unimaginable. Though such progress is never as simple as it appears, at least, in this case, it has appeared so.

Assistant Chief Mario Chapple

Assistant Chief Mario Chapple & Former Chief Robert C. Allison, Jr

ABOUT THE AUTHORS

IAN SPANIER began taking photographs at six years old when his parents gave him his first point and shoot camera. After majoring in photography in college, Spanier worked in publishing as an editor, but making pictures never left him. Having only known 35mm, he taught himself medium and large format as well as lighting.

Ian's work has won numerous awards and accolades from the photographic community for his commercial work. Clients include *MTV, A&E, HBO, Danskin, Conde Nast Traveler, Shape, Muscle & Fitness, Men's Fitness, Runner's World, Marie Claire, Time Out NY, The New York Times, Los Angeles Magazine, AirTran's Go Magazine, The New York Economic Development Corporation, Bank of America, Gerber Knife Company, Time Inc.,* and *Simon and Schuster,* among many others. His most current work can be seen at www.ianspanier.com.

Ian's first full book of published work, *Playboy, a Guide to Cigars,* arrived in cigar shops in November 2009 and the public version hit retail stores Spring 2010. The book is a collection of his photographs made in six countries spanning two and a half years.

The original masters of photography have always inspired Spanier as they shot what they saw. For him, there is no "one" subject that he photographs; he also chooses to shoot what he sees.

MAREK FUCHS is a writer, teacher, and volunteer firefighter. He is on the non-fiction writing faculty at Sarah Lawrence College, wrote *The New York Times* "County Lines" column for six years, and is the author of *A Cold-Blooded Business,* called "riveting" by Kirkus Reviews. Fuchs the firefighter advises all to check their smoke alarm batteries tonight. He can be reached through Twitter at @MarekFuchs.

GRACE MARTINEZ is a Senior Associate Art Director at *Glamour Magazine.* A graduate from Fordham University Lincoln Center, Grace has worked for a number of national publications including *Conde Nast Portfolio, Women's Health,* and *Latina* magazine. Her design work has been recognized by *Print Magazine* and the Society of Publication Designers. She currently resides in Brooklyn, New York.

FLORIAN BACHLEDA is the Creative Director of *Fast Company.* Previously, he was the Creative Director for *Latina* and Design Director for *Vibe, Vibe Vixe*n, *Maximum Golf,* and *P.O.V. (Point of View).* He has also worked at *New York Magazine, Entertainment Weekly,* and *The Village Voice.* Since 2006, he has been the Creative Director of FB Design, whose clients included Time Inc., Condé Nast, and Hearst, among others. The studio now works solely on personal projects.

Bachleda has won awards and medals from the Society of Publication Designers, American Illustration, American Photography, Print Magazine and Communication Arts, among others. He has taught at the School of Visual Arts and has chaired and juried numerous design, photography, and illustration competitions. He has served as President of the Society of Publication Designers and was on its Board of Directors for five years.